The
Monte Carlo
Method

Popular Lectures in Mathematics

*Survey of Recent East European Mathematical
Literature*

A project conducted by
IZAAK WIRSZUP,
Department of Mathematics,
the University of Chicago,
under a grant from the
National Science Foundation

I. M. Sobol'

The Monte Carlo Method

Translated and adapted from the second Russian edition by Robert Messer, John Stone, and Peter Fortini

The University of Chicago Press
Chicago and London

The University of Chicago Press, Chicago 60637
The University of Chicago Press, Ltd., London

International Standard Book Number: 0–226–76749–3
Library of Congress Catalog Card Number: 73–89791

I. M. SOBOL' is a research mathematician
at the Mathematical Institute of the
USSR Academy of Sciences.
[1974]

Contents

Preface

Some years ago I accepted an invitation from the Department of Computer Technology at the Public University to deliver two lectures on the Monte Carlo method. These lectures have since been repeated over the course of several years and their contents have gradually settled and "jelled." The present edition also includes a supplementary section (chapter 2), about which I should say a few words.

Shortly before the first lecture, I discovered to my horror that most of the audience was unfamiliar with probability theory. Since some familiarity with that theory was absolutely necessary, I hurriedly inserted in the lecture a section acquainting my listeners with some basic concepts of probability. Chapter 2 of this booklet is an outgrowth of that section.

Surely everyone has heard and used the words "probability," "frequency," and "random variable." The intuitive notions of probability and frequency more or less correspond to the true meanings of the terms, but the layman's notion of a random variable is rather different from the mathematical definition. In chapter 2, therefore, the concept of probability is assumed to be known, and only the more complex concept of a random variable is explained at length. This section cannot replace a course in probability theory: the presentation here is greatly simplified, and no proofs are given of the theorems asserted. But it does give the reader enough acquaintance with random variables for an understanding of the simplest procedures of the Monte Carlo method.

The principal goal of this booklet is to suggest to specialists in all areas that they will encounter problems which can be solved by the Monte Carlo method.

The problems considered in the lectures are fairly simple and have been drawn from diverse fields. Naturally, they cannot encompass all the areas in which the method can be applied. For example, there is not a word about medicine in the booklet, although the methods of chapter 7 do enable one to calculate radiation dosages in X-ray therapy. If one has a program for calculating the absorption of radiation by the various body tissues, it is possible to select the dosage and direction of irradiation which most effectively ensures that no harm is done to healthy tissues.

The present book includes the material read in the lectures. A more detailed exposition is given of certain examples, and chapter 9 has been added.

I. Sobol'
Moscow, 1967

1

Introduction to the Method

The Monte Carlo method is a method of approximately solving mathematical and physical problems by the simulation of random quantities.

1.1. The Origin of the Monte Carlo Method

The generally accepted birth date of the Monte Carlo method is 1949, when an article entitled "The Monte Carlo Method"[1] appeared. The American mathematicians J. Neyman and S. Ulam are considered its originators. In the Soviet Union, the first articles on the Monte Carlo method were published in 1955 and 1956.[2]

The theoretical basis of the method has long been known. In the nineteenth and early twentieth centuries, statistical problems were sometimes solved with the help of random selections, that is, in fact, by the Monte Carlo method. Prior to the appearance of electronic computers, this method was not widely applicable since the simulation of random quantities by hand is a very laborious process. Thus, the beginning of the Monte Carlo method as a highly universal numerical technique became possible only with the appearance of computers.

The name "Monte Carlo" comes from the city of Monte Carlo in the principality of Monaco, famous for its gambling house. One of the simplest mechanical devices for obtaining random quantities is the roulette wheel. This subject will be considered in chapter 3. Perhaps it is worthwhile to answer here the frequently asked question: "Does the

1. N. Metropolis and S. Ulam, "The Monte Carlo Method," *Journal of the American Statistical Association* 44, no. 247 (1949):335–41.
2. These were the articles by V. V. Chavchanidze, Yu. A. Schreider, and V. S. Vladimirov.

Monte Carlo method help one win at roulette?" The answer is that it does not; it is not even an attempt to do so.

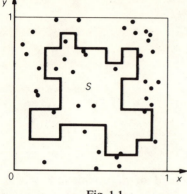

Fig. 1.1

Example. In order to make more clear to the reader what we are talking about, let us examine a very simple example. Suppose that we need to compute the area of a plane figure S. This may be a completely arbitrary figure with a curvilinear boundary, given graphically or analytically, connected or consisting of several pieces. Let the region be as represented in figure 1.1, and let us assume that it is contained completely within the unit square.

Choose at random N points in the square and designate the number of points lying inside S by N'. It is geometrically obvious that the area of S is approximately equal to the ratio N'/N. The greater the N, the greater the accuracy of this estimate.

In the example represented in figure 1.1, we selected $N = 40$ points. Of these, $N' = 12$ points appeared inside S. The ratio $N'/N = 12/40 = 0.30$, while the true area of S is 0.35.[3]

1.2. Two Features of the Monte Carlo Method

In our example it would not have been too difficult to calculate directly the true area of S. In Part II of this book we shall consider some less trivial examples. Our simple method, however, does point out one feature of the Monte Carlo method, that is, the simple structure of the computational algorithm. This algorithm consists, in general, of a process for producing a random event. The process is repeated N times, each trial being independent of the rest, and the results of all the trials are averaged together. Because of its similarity to the process of performing a scientific experiment, the Monte Carlo method is sometimes

3. In practice, the Monte Carlo method is not used for calculating the area of a plane figure. There are other methods for this, which, although they are more complicated, guarantee much greater accuracy.

Still, the Monte Carlo method shown in our example permits us to calculate very simply the "many-dimensional volume" of a body in many-dimensional space; and in such a case the Monte Carlo method often proves to be the only numerical method useful in solving the problem.

called the *method of statistical trials*. In our example, the random event consisted of taking a random point in the square and checking to determine whether it belonged to S, and the results of the trials were averaged together by taking the ratio N'/N.

A second feature of the method is that, as a rule, the error which we expect from the calculation is $\sqrt{(D/N)}$, where D is some constant and N the number of trials. In our example, it turns out from probability theory (for proof, see section 2.6) that

$$D = A(1 - A) = (0.35)(1 - 0.35) \approx 0.23 ,$$

where A is the true area of the region S, so $\sqrt{(D/N)} = \sqrt{(0.23/40)} \approx 0.076$. We see that the actual error of the calculation, 0.05, was not, after all, unreasonably large.

From the formula

$$\text{error} \approx \sqrt{\left(\frac{D}{N}\right)}$$

it is clear that to decrease the error by a factor of 10 (in other words, to obtain another significant digit in the result), it is necessary to increase N (and the amount of work) by a factor of 100.

To attain high precision in this way is clearly impossible. The Monte Carlo method is most effective in solving problems in which the result need be accurate only to 5–10%. However, any particular problem can be solved by different variations of the Monte Carlo method[4] which assign different values to D. In many problems, a computational procedure which gives D a significantly smaller value will considerably increase the accuracy of the result.

1.3. Problems That Can Be Solved by the Monte Carlo Method

The Monte Carlo method makes possible the simulation of any process influenced by random factors. This, however, is not its only use. For many mathematical problems involving no chance, we can artificially devise a probabilistic model (frequently several) for solving these problems. In fact, this was done in the example in section 1.1. For these reasons the Monte Carlo method can be considered a universal method for solving mathematical problems.

4. In foreign literature the term Monte Carlo *methods* (in the plural) is now more frequently used, in view of the fact that the same problem can be solved by simulating different random variables.

It is particularly interesting that in certain cases, instead of simulating the actual random process, it is advantageous to use artificial models. Such a situation is the topic of chapter 7.

More about the example. Let us return to the example of section 1.1. For the calculation we needed to choose points at random in the unit square. How is this actually done?

Let us set up such an experiment. Imagine figure 1.1 (on an increased scale) hanging on a wall as a target. Some distance from the wall, N darts are aimed at the center of the square and thrown. Of course, not all the darts will fall exactly in the center; they will strike the target at N random points.[5] Can these points be used to estimate the area of S?

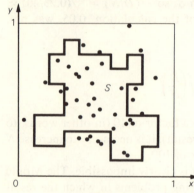

The result of such an experiment is depicted in figure 1.2. In this experiment $N = 40$, $N' = 24$, and the ratio $N'/N = 0.60$ is almost double the true value of the area (0.35). It is clear that when the darts are thrown with very great skill, the result of the experiment will be very bad, as almost all of the darts will fall near the center and thus in S.[6]

Fig. 1.2

We can see that our method of computing the area will be valid only when the random points are not "simply random," but, in addition, "uniformly distributed" over the whole square. To give these words a precise meaning, we must become acquainted with the definition of random variables and with some of their properties. This information is presented in chapter 2. A reader who has studied probability theory may omit all except sections 2.5 and 2.6 of chapter 2.

5. We assume that the darts are not in the hands of the world champion and that they are thrown from a sufficiently great distance from the target.

6. The ways in which the random points were chosen in figures 1.1 and 1.2 will be discussed in section 4.5.

Part 1

**Simulating
Random Variables**

2

Random Variables

We assume that the concept of *probability* is more or less familiar to the reader, and we pass directly to the concept of a *random variable*.

The words "random variable," in ordinary English usage, refer to the outcome of any process which proceeds without any discernible aim or direction. However, a mathematician's use of the words "random variable" has a completely definite meaning. He is saying that we do not know the value of this quantity in any given case, but we know what values it can assume and we know the probabilities with which it assumes these values. On the basis of this information, while we cannot precisely predict the result of any single trial associated with this random variable, we can predict very reliably the total results of a great number of trials. The more trials there are (as they say, the larger the sample is), the more accurate the prediction will be.

2.1. Discrete Random Variables

The random variable X is called *discrete* if it can assume any of a discrete set of values $x_1, x_2, \ldots, x_n$.[1]

X is therefore defined by the table

$$X = \begin{pmatrix} x_1 & x_2 & \cdots & x_n \\ p_1 & p_2 & \cdots & p_n \end{pmatrix}, \tag{T}$$

where $x_1, x_2, \ldots, x_n$ are the possible values of the variable X, and $p_1, p_2, \ldots, p_n$ are the probabilities corresponding to them. Precisely

---

1. In probability theory discrete random variables that can assume a countably infinite number of values $x_1, x_2, x_3, \ldots$ are also considered.

speaking, the probability that the random variable has the value x_i (denoted by $P(X = x_i)$) is equal to p_i:

$$P(X = x_i) = p_i .$$

Sometimes we write $p_X(x_i)$ instead of p_i or $P(X = x_i)$.

Table (T) is called the *distribution of the random variable*.

The numbers $x_1, x_2, \ldots, x_n$ are arbitrary. However, the probabilities $p_1, p_2, \ldots, p_n$ must satisfy two conditions:

(a) all p_i are non-negative:

$$p_i \geq 0 ; \tag{2.1}$$

(b) the sum of all the p_i equals 1:

$$p_1 + p_2 + \cdots + p_n = 1 . \tag{2.2}$$

The last condition means that in every event X *must* assume one of the values $x_1, x_2, \ldots, x_n$.

The number

$$E(X) = \sum_{i=1}^{n} x_i p_i \tag{2.3}$$

is called the *expected value*, or *mathematical expectation*, of the random variable X.

To illustrate the physical meaning of this quantity we write it in the following form:

$$E(X) = \frac{\sum_{i=1}^{n} x_i p_i}{\sum_{i=1}^{n} p_i} .$$

We see that $E(X)$ is in a sense the *average value* of the variable X, in which the more probable values are added into the sum with larger weights.[2]

---

2. Averaging with weights is very common in science. For example, in mechanics; if masses $m_1, m_2, \ldots, m_n$ are distributed on the x-axis at the points $x_1, x_2, \ldots, x_n$, then the abscissa of the center of gravity of this system is given by the formula

$$\bar{X} = \frac{\sum_{i=1}^{n} x_i m_i}{\sum_{i=1}^{n} m_i} .$$

Of course, in this case the sum of all the masses does not necessarily equal unity.

Let us mention the basic properties of mathematical expectation. If c is any constant, then

$$E(X + c) = E(X) + c, \qquad (2.4)$$

$$E(cX) = cE(X). \qquad (2.5)$$

If X and Y are any two random variables, then

$$E(X + Y) = E(X) + E(Y). \qquad (2.6)$$

The number

$$\text{Var}(X) = E((X - E(X))^2) \qquad (2.7)$$

is called the *variance* of the random variable X. That is, the variance Var (X) is the mathematical expectation of the squared deviation of the random variable X from its average value $E(X)$. Obviously, Var $(X) \geq 0$ always.

The mathematical expectation and the variance are the most important numbers characterizing the random variable X. What is their practical value?

If we observe the variable X many times and obtain the values $X_1, X_2, \ldots, X_N$ (each of which is equal to one of the numbers $x_1, x_2, \ldots, x_n$), then the arithmetic mean of these numbers will be close to $E(X)$:

$$\frac{1}{N}(X_1 + X_2 + \cdots + X_N) \approx E(X); \qquad (2.8)$$

and the variance Var (X) characterizes the spread of these values around the average $E(X)$.

Formula (2.7) can be transformed using formulas (2.4)–(2.6):

$$\text{Var}(X) = E(X^2 - 2E(X)\cdot X + (E(X))^2)$$
$$= E(X^2) - 2E(X)\cdot E(X) + (E(X))^2,$$

whence

$$\text{Var}(X) = E(X^2) - (E(X))^2. \qquad (2.9)$$

It is usually easier in hand computations to find the variance by formula (2.9) than by formula (2.7).

Let us mention the basic properties of the variance: If c is any constant, then

$$\text{Var}\,(X + c) = \text{Var}\,(X)\,, \qquad (2.10)$$

$$\text{Var}\,(cX) = c^2\,\text{Var}\,(X)\,. \qquad (2.11)$$

The concept of *independence* of random variables plays an important role in the theory of probability. Let us suppose that, besides the variable X, we also watch a random variable Y. If the distribution of the variable X does not change when we know the value which the variable Y assumes, and vice versa, then it is natural to believe that X and Y do not depend on each other. We then say that the random variables X and Y are independent.

The following relations hold for independent random variables X and Y:

$$E(XY) = E(X)E(Y)\,, \qquad (2.12)$$

$$\text{Var}\,(X + Y) = \text{Var}\,(X) + \text{Var}\,(Y)\,. \qquad (2.13)$$

Example. Let us consider a random variable X with the distribution

$$X = \begin{pmatrix} 1 & 2 & 3 & 4 & 5 & 6 \\ \frac{1}{6} & \frac{1}{6} & \frac{1}{6} & \frac{1}{6} & \frac{1}{6} & \frac{1}{6} \end{pmatrix}.$$

Since each of the values is equally probable, the number of dots appearing when a die is thrown can be used to realize these values. Let us calculate the mathematical expectation and the variance of X. By formula (2.3),

$$E(X) = \tfrac{1}{6}(1 + 2 + 3 + 4 + 5 + 6) = 3.5\,.$$

By formula (2.9),

$$\text{Var}\,(X) = E(X^2) - (E(X))^2$$

$$= \tfrac{1}{6}(1^2 + 2^2 + 3^2 + 4^2 + 5^2 + 6^2) - (3.5)^2 = 2.917\,.$$

Example. Let us consider the random variable Y with distribution

$$Y = \begin{pmatrix} 3 & 4 \\ \frac{1}{2} & \frac{1}{2} \end{pmatrix}.$$

To realize these values, we can consider a toss of a coin with the condition that a head counts 3 points and a tail 4 points. In this case,

$$E(Y) = 0.5 \cdot 3 + 0.5 \cdot 4 = 3.5\,;$$

$$\text{Var}\,(Y) = 0.5(3^2 + 4^2) - (3.5)^2 = 0.25\,.$$

We see that $E(Y) = E(X)$, but Var $(Y) <$ Var (X). This could easily have been anticipated, since the values of Y can differ from 3.5 only by ± 0.5, while for the values of X the spread can reach ± 2.5.

2.2. Continuous Random Variables

Let us assume that some radium is placed at the origin of a coordinate plane. As each atom of radium decays, an α-particle is emitted. We shall describe its direction by the angle ψ (fig. 3). Since both in theory and practice any direction of flight is possible, this random variable can assume any value from 0 to 2π.

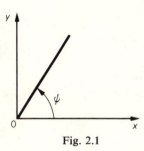

Fig. 2.1

We shall say that a random variable X is *continuous* if it can assume any value in some interval $[a, b]$.

A continuous random variable X is defined by the assignment of a function $p(x)$ to the interval $[a, b]$ containing the possible values of this variable. $p(x)$ is called the *probability density* or *density distribution* of the random variable X.

The significance of $p(x)$ is as follows: Let (a', b') be an arbitrary interval contained in $[a, b]$ (that is, $a \leq a'$, $b' \leq b$). Then the probability that X lies in the interval (a', b') is equal to the integral

$$P(a' < X < b') = \int_{a'}^{b'} p(x)\, dx\,. \qquad (2.14)$$

In figure 2.2 the shaded area represents the value of the integral (2.14).

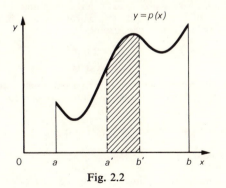

Fig. 2.2

The set of values of X can be any interval. The interval may contain either or both of its endpoints, and even the cases $a = -\infty$ and $b = \infty$ are possible. The density $p(x)$, however, must satisfy two conditions analogous to conditions (1) and (2) for discrete variables:

(a) the density $p(x)$ is nonnegative:

$$p(x) \geq 0. \tag{2.15}$$

(b) the integral of the density $p(x)$ over the whole interval (a, b) is equal to 1:

$$\int_a^b p(x)\, dx = 1. \tag{2.16}$$

The number

$$E(X) = \int_a^b xp(x)\, dx \tag{2.17}$$

is called the *expected value* of a continuous random variable.

The significance of this quantity is the same as in the case of the discrete random variable. Indeed, since

$$E(X) = \frac{\int_a^b xp(x)\, dx}{\int_a^b p(x)\, dx},$$

it is easily seen that this is the average value of X. In fact, X can assume any value x in the interval (a, b) with "weight" $p(x)$.[3]

Everything explained in section 2.1 from formula (2.4) up to and including formula (2.13) is also valid for continuous random variables. This includes the definition of variance (2.7), the formula (2.9) for its computation, and all the properties of $E(X)$ and Var (X). We shall not repeat them.[4]

---

3. In this case it is also possible to explain the analogous formula in mechanics: If the linear density of a rod is equal to $p(x)$ for $a \leq x \leq b$, then the abscissa of the center of gravity is given by the formula

$$\overline{X} = \frac{\int_a^b xp(x)\, dx}{\int_a^b p(x)\, dx}.$$

4. This statement is not exactly true for all continuous random variables. In statistics there arise a few continuous random variables for which one or both of the integrals

$$E(X) = \int xp(x)\, dx, \qquad \text{Var } (X) = \int x^2 p(x)\, dx - (E(X))^2$$

diverge; for instance, the *Cauchy density* $p(x) = (1/\pi)(1/[1 + x^2])$, for $-\infty < x < \infty$, has infinite variance. For these variables, formulas (2.7) through (2.13) cannot be used, and special methods must be devised to treat them.

Let us mention just one more formula, that for the mathematical expectation of a random function. As before, let the random variable X have probability density $p(x)$. We choose an arbitrary continuous function $f(x)$, and consider the random variable $Y = f(X)$, sometimes called a *random function*. It can be proved that

$$E(f(X)) = \int_a^b f(x)p(x)\,dx\,. \tag{2.18}$$

Let us stress that, generally speaking, $E(f(X)) \neq f(E(X))$.

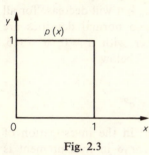

Fig. 2.3

The random variable G defined on the interval $[0, 1]$ and having a density $p(x) = 1$ is called a uniform distribution on $[0, 1]$ (fig. 2.3).

Whatever subinterval (a', b') we take within $[0, 1]$, the probability that G lies in (a', b') is equal to

$$\int_{a'}^{b'} p(x)\,dx = b' - a'\,,$$

that is, the length of the subinterval. In particular, if we divide $[0, 1]$ into any number of intervals of equal length, the probabilities of G hitting any of these intervals are the same.

It is easy to calculate that

$$E(G) = \int_0^1 xp(x)\,dx = \int_0^1 x\,dx = \tfrac{1}{2}\,,$$

$$\mathrm{Var}\,(G) = \int_0^1 x^2 p(x)\,dx - (E(G))^2 = \tfrac{1}{3} - \tfrac{1}{4} = \tfrac{1}{12}\,.$$

In what follows we shall have many uses for the random variable G.

2.3. Normal Random Variables

A *normal* (or *gaussian*) random variable is a random variable Z defined on the whole axis $(-\infty, \infty)$ and having the density

$$p(x) = \frac{1}{\sigma\sqrt{2\pi}} \exp\left[-\frac{(x-a)^2}{2\sigma^2}\right]\,, \tag{2.19}$$

where a and $\sigma > 0$ are numerical parameters.

The parameter a does not affect the shape of the curve $p(x)$: a change in a results only in a displacement of the curve along the x-axis. However, the shape of the curve does change with a change in σ. Indeed, it is easy to see that

$$\max (p(x)) = p(a) = \frac{1}{\sigma\sqrt{2\pi}} :$$

If σ decreases, the $\max (p(x))$ will increase. However, according to condition (2.16), all the area under the curve $p(x)$ is equal to 1. Therefore, the curve will extend upward near $x = a$, but will decrease for all sufficiently large values of x. In figure 6 two normal densities are drawn, one with $a = 0$, $\sigma = 1$, and another with $a = 0$, $\sigma = 0.5$. (Another normal density is drawn in figure 6.5 below.)

It is possible to show that

$$E(Z) = a , \qquad \mathrm{Var}\,(Z) = \sigma^2 .$$

Normal random variables are encountered in the investigation of very diverse problems. For example, an error δ in measurement is generally a normal random variable. The reason for this will be discussed shortly. If the error in measurement is not systematic, then $a = E(\delta) = 0$. And the quantity $\sigma = \sqrt{\mathrm{Var}\,(\delta)}$, called the *standard deviation* of δ, describes the error in the method of measurement.

The rule of "three sigmas." It is not difficult to determine that for a normal density p,

$$\int_{a+3\sigma}^{a+3\sigma} p(x)\,dx = 0.997 ,$$

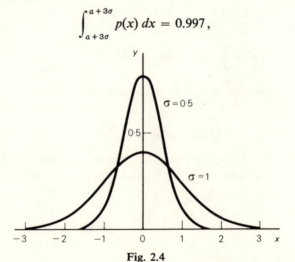

Fig. 2.4

whatever a and σ are in (2.19). From (2.14) it follows that

$$P(a - 3\sigma < Z < a + 3\sigma) = 0.997. \tag{2.20}$$

The probability 0.997 is very near to 1. We therefore give the latter formula the following interpretation: *For a single trial it is practically impossible to obtain a value Z differing from E(Z) by more than 3σ.*

2.4. The Central Limit Theorem of Probability Theory

This remarkable theorem was first formulated by Laplace. Many mathematicians, including P. L. Chebyshev, A. A. Markov, and A. M. Lyapunov, have worked on generalizations of the original result. Its proof is rather complex.

Let us consider N independent, identically distributed random variables $X_1, X_2, \ldots, X_N$; that is to say, the probability densities of these variables coincide. Consequently, their mathematical expectations and variances also coincide.

We write

$$E(X_1) = E(X_2) = \cdots = E(X_N) = m,$$

$$\text{Var}(X_1) = \text{Var}(X_2) = \cdots = \text{Var}(X_N) = v^2.$$

Denote the sum of all these variables by S_N:

$$S_N = X_1 + X_2 + \cdots + X_N.$$

From formulas (2.6) and (2.13) it follows that

$$E(S_N) = E(X_1 + X_2 + \cdots + X_N) = Nm,$$

$$\text{Var}(S_N) = \text{Var}(X_1 + X_2 + \cdots + X_N) = Nv^2.$$

Now let us consider the normal random variable Z_N with these same parameters: $a = Nm$, $\sigma^2 = Nv^2$.

THEOREM 2.1. *The density of the sum S_N approaches the density of the normal variable Z_N in such a way that for every x,*

$$p\left(\frac{S_N - Nm}{v\sqrt{(N)}} < x\right) \approx p\left(\frac{Z_N - Nm}{v\sqrt{(N)}} < x\right)$$

for all large N.

The significance of this theorem is clear: The sum S_N of a large number of identical random variables is approximately normal $(p_{S_N}(x) \approx p_{Z_N}(x))$.

Indeed, the theorem is valid under considerably weaker conditions. Not all the terms $X_1, X_2, \ldots, X_N$ have to be identical and independent; essentially, all that is required is that single terms do not play too great a role in the sum.

It is precisely this theorem which explains why normal random variables are so often encountered in nature. Indeed, whenever we meet a summing influence over a large number of independent random factors, the resulting random variable proves to be normal. For example, the scattering of artillery shells from their target is almost always a normal random variable, since it depends on the meteorological conditions in all the various regions of the trajectory as well as on many other factors.

2.5. The General Scheme of the Monte Carlo Method

Suppose that we want to determine some unknown quantity m. Let us attempt to devise a random variable X with $E(X) = m$. Say the variance of this variable is Var $(X) = v^2$.

Consider N independent random variables $X_1, X_2, \ldots, X_N$, with distributions identical to that of X. If N is sufficiently large, then, according to the theorem of section 2.4, the distribution of the sum $S_N = X_1 + X_2 + \cdots + X_N$ will be approximately normal with parameters $a = Nm$, $\sigma^2 = Nv^2$. From equation (2.20) it follows that

$$P(Nm - 3v\sqrt{N} < S_N < Nm + 3v\sqrt{N}) \approx 0.997.$$

If we divide the inequality within the parentheses by N, we obtain an equivalent inequality, and the probability remains the same:

$$P\left(m - \frac{3v}{\sqrt{N}} < \frac{S_N}{N} < m + \frac{3v}{\sqrt{N}}\right) \approx 0.997.$$

We can rewrite the last relation in a slightly different form:

$$P\left(\left|\frac{1}{N}\sum_{j=1}^{N} X_j - m\right| < \frac{3v}{\sqrt{N}}\right) \approx 0.997. \tag{2.21}$$

This is an extremely important relation for the Monte Carlo method. It gives us both a method of calculating m and an estimate of the uncertainty of our estimation.

Indeed, suppose that we have found N values of the random variable

X.[5] From (2.21) it is obvious that the arithmetic mean of these values will be approximately equal to m. With high probability, the error of this approximation does not exceed the quantity $3v/\sqrt{N}$. Obviously, this error converges to zero as N increases.

2.6. A Last Word about the Example

Let us now apply some of these ideas to the example of section 1.1 to see how we originally obtained the formula for error

$$\sqrt{\left(\frac{D}{N}\right)} = \sqrt{\left(\frac{A(1 - A)}{N}\right)} \, .$$

If we denote the result of the jth single trial by

$$X_j = \begin{cases} 1, & \text{if the } j\text{th random point lies in } S \\ 0, & \text{if not}, \end{cases}$$

then our estimate of the area of S is just $\sum X_j/N$. It is easy to see that the distribution of each X_j is

$$\begin{pmatrix} 0 & 1 \\ 1 - A & A \end{pmatrix} .$$

Hence, by formulas (2.3) and (2.9),

$$m = E(X) = 0 \cdot (1 - A) + 1 \cdot A = A \, ,$$

$$v^2 = \text{Var}(X) = 0^2 \cdot (1 - A) + 1^2 \cdot A - A^2 = A(1 - A) \, ,$$

$$\frac{v}{\sqrt{N}} = \sqrt{\left(\frac{D}{N}\right)} = \sqrt{\left(\frac{A(1 - A)}{N}\right)} \, .$$

We have chosen to omit the factor 3 from the formula $3v/\sqrt{N}$ since a deviation so large as $3(v/\sqrt{N})$ will rarely be encountered. Our formula $\sqrt{(D/N)}$ actually gives the standard deviation of the normal random variable which is "closest" to the distribution of $(\sum X_j)/N$. It is closely related to another measure of error, the *probable error*, which we shall introduce later, in chapter 8.

5. It is immaterial whether we find one value of each of the variables X_1, $X_2, \ldots, X_N$ or N values for the single variable X, since all the random variables have identical distributions.

3

Generating
Random Numbers
on a Computer

The very thought of generating random numbers on a computer sometimes provokes the question: "If everything the machine does must be programmed beforehand, where can randomness come from?" There are, indeed, several difficulties associated with this point, but they belong more to philosophy than to mathematics, and so we shall not dwell on them.

The random variables discussed in chapter 2 are ideal mathematical concepts. The question is whether one can use them to describe natural phenomena experimentally. Such a description, of course, always proves to be approximate, and a random variable which describes some physical quantity with perfect accuracy in one set of phenomena can prove to characterize the same quantity poorly during the investigation of others.

Such problems of description are universal not only within applied mathematics but in all other fields as well. A cartographer, for example, can draw a road on a national map as a perfectly straight line. On the large-scale map of a heavily populated area, however, it must be drawn wide and crooked, and very close examination reveals all sorts of properties of the road: color, texture, and the like, of which the original description can take no account whatsoever. Our use of random variables should be regarded not as providing a perfect description of natural phenomena, but as a tool in solving particular problems in which we may be interested.

Ordinarily, three ways of obtaining random values are distinguished: tables of random numbers, random number generators, and the pseudo-random number method.

3.1. Tables of Random Numbers

Let us perform the following experiment. We mark the digits 0, 1, 2, ..., 9 on ten identical slips of paper. We place these slips of paper in a hat, mix them together, and take out one; then return it and mix again. We write down the digits obtained in this way in the form of a table like table A in the Appendix (in table A the digits are arranged in groups of five for convenience).

Such a table is called a *table of random digits*. It is possible to put it into a computer's memory. Then, in the process of calculation, when we need values of a random variable with the distribution

$$\begin{pmatrix} 0 & 1 & 2 & \cdots & 9 \\ 0.1 & 0.1 & 0.1 & \cdots & 0.1 \end{pmatrix}, \tag{3.1}$$

we need only take the next digit from this table.

The largest of the published random-number tables contains one million digits.[1] Of course, it was compiled with the assistance of technical equipment more sophisticated than a hat: A special roulette wheel was constructed which operated electronically. Figure 3.1 shows an elementary version of such a wheel. A rotating disc is stopped suddenly, and the number to which the stationary arrow points is selected.

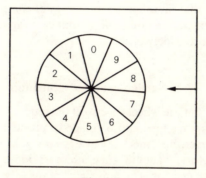

Fig. 3.1

Compiling a good table of random numbers is not as easy as it may appear. Any real physical device produces random variables with a distribution differing slightly from the ideal distribution. During an experiment there may well be accidents (for example, one of the slips of paper in the hat might stick to the lining for some time). Therefore, the compiled tables are carefully checked by special statistical tests, to make sure that no particular characteristics of the group of numbers

1. RAND Corporation, *A Million Random Digits with 100,000 Normal Deviates* (Glencoe: Free Press, 1955).

contradict the hypothesis that the numbers are independent values of a random variable (3.1).

Let us examine one of the simplest tests. Consider a table containing N digits. Let the number of zeros in this table be v_0, the number of ones v_1, the number of twos v_2, and so on. We calculate the sum

$$\sum_{i=0}^{9} (v_i - (0.1)N)^2 .$$

The theory of probability allows us to predict the range in which this sum should lie. It should not be very large, since the mathematical expectation of each of the v_i is equal to $(0.1)N$, but neither should it be too small, since that would indicate an "overly regular" distribution of values.

Tables of random numbers are used only for Monte Carlo method calculations performed by hand. The fact is that all computers have comparatively small internal memories, and a large table will not fit in them. To store the table in external memory and then to consult it continually for numbers slows calculation considerably.

The possibility that, in time, the memories of computers will increase sharply should not be ruled out, and in that case random-number tables might become more widely useful.

3.2. Random-Number Generators

It would seem that the wheel described in section 3.1 could be hooked up to a calculating machine and be made to produce random numbers as needed. However, any mechanical device would be too slow for a computer. Therefore, vacuum tube noise is more often used as a random-number generator. The noise level of the tube is monitored, and if, within some fixed interval of time, the noise exceeds a set threshold an even number of times, a zero is recorded; if an odd number of times, a one.[2]

At first glance this is a very convenient procedure. Suppose m such generators work in parallel, all the time, and send random zeros and ones into all the binary places of a particular memory location. At any point in its calculations the machine can go to this location and take from it the random value G. The values will be evenly distributed over the interval [0, 1], though only approximately, of course, each number

2. There are setups which are even more statistically perfect.

being an m-digit binary fraction of the form 0. $D_{(1)}D_{(2)} \ldots D_{(m)}$, where each of the variables $D_{(i)}$ imitates a random variable with the distribution

$$\begin{pmatrix} 0 & 1 \\ \frac{1}{2} & \frac{1}{2} \end{pmatrix}.$$

Yet even this method is not free from defects. First, it is difficult to check the "quality" of the numbers produced. It is necessary to make periodic tests, since any imperfection can lead to a "distribution drift" (that is, the zeros and ones in one of the places begin to appear in unequal frequencies). Second, it is often desirable to be able to repeat a calculation on the computer. But it is impossible to duplicate a sequence of random numbers if they are not held in the memory throughout the calculation; and if they are held in the memory, we are back to the random-number tables.

Methods of this sort will undoubtedly prove useful when computers are constructed especially for solving problems by means of the Monte Carlo method. For all-purpose computers, however, on which calculations requiring random numbers come up only rarely, it is simply not economical to maintain and to make use of such special equipment. It is better to use pseudo-random numbers.

3.3. Pseudo-Random Numbers

So long as the "quality" of the random numbers used can be verified by special tests, one can ignore the means by which they were produced. It is even possible to try to generate them through a set formula.

Numbers obtained by a formula that imitate the values of a random variable G uniformly distributed in $[0, 1]$ are called *pseudo-random numbers*. Here the word "imitate" means that these numbers satisfy the test just as if they were values of a random variable. They will be quite satisfactory so long as the calculations performed with them remain unrelated to the particular formula by which they were produced.

The first algorithm for obtaining pseudo-random numbers was proposed by J. Neyman. It is called the *middle-of-squares method*. We illustrate it with an example.

We are given a four-digit integer $n_1 = 9876$. We square it. We usually obtain an eight-digit number $n_1^2 = 97535376$. We take out the middle four digits of this number and designate the result $n_2 = 5353$.

Then we square n_2 ($n_2^2 = 28654609$) and once more take out the middle four digits, obtaining $n_3 = 6546$.

Then $n_3{}^2 = 42850116$, $n_4 = 8501$; $n_4{}^2 = 72267001$, $n_5 = 2670$; $n_5{}^2 = 07128900$, $n_6 = 1289$, and so forth.

The proposed values to be used for the variable G are then 0.9876; 0.5353; 0.6546; 0.8501; 0.2670; 0.1289, and so forth.[3]

This algorithm is unfortunately not suitable, for it tends to give more small numbers than it should. It is also prone to falling into "traps," such as the sequences 0000, 0000, ..., and 6100, 2100, 4100, 8100, 6100, ... For these reasons various experimenters have worked out other algorithms. Some of them take advantage of peculiarities of specific computers. As an example, let us examine one such algorithm, used on the Strela computer.

Example.[4] The Strela is a triple-address, floating-point computer. The memory location into which the number x is placed is made up of forty-three binary places (fig. 3.2). The machine works with binary

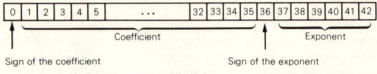

Fig. 3.2

numbers in the form $x = \pm q \cdot 2^{\pm p}$, where p is the exponent of the number and q the coefficient.[5] In the jth place there can be a zero or a one; let us call this value e_j. Then

$$q = \frac{e_1}{2^1} + \frac{e_2}{2^2} + \cdots + \frac{e_{35}}{2^{35}}, \qquad p = e_{37}2^5 + e_{38}2^4 + \cdots + e_{42}2^0.$$

In locations 0 and 36, zero represents the $+$ sign, one the $-$ sign.

3. This algorithm can be written in the form $n_{k+1} = F(n_k)$, where F stands for the aggregate of the operations that are performed on the number n_k in order to obtain n_{k+1}. The number n_1 is given. The pseudo-random numbers $G_k = 10^{-4}n_k$.

4. See I. M. Sobol', "Psevdosluchainye chisla dlya mashiny *Strela*" [Pseudo-random numbers for the Strela computer], *Teoriya veroyatnosti i ee primeneniya* [Probability theory and its applications] 3, no. 2 (1958):205–11.

5. A somewhat different method of floating-point number storage is common in American computers such as the IBM 360. Either 32 or 64 binary places, arranged in groups of eight, are used. Real numbers are considered as written in the form

$$\pm q \cdot (16)^{p-64},$$

where $1/16 \le q < 1$, $0 \le p \le 127$. The leftmost binary digit records the sign of the number, 0 for $+$, 1 for $-$; the next seven places record the value of p written as a binary integer; the last 24 or 56 places give the value of the coefficient q. A method of generating random numbers similar to that of the author can easily be developed for use with this arrangement.—*Trans.*

After a nonzero number G_0, usually 1, is chosen, the number G_{k+1} is obtained from G_k in three operations:

(1) G_k is multiplied by a large constant, usually 10^{17}.

(2) The representation of the product $10^{17}G_k$ is displaced seven places to the left, so that the first seven places of the product disappear, and zeros appear in places 36 to 42.

(3) The absolute value of the resulting number is taken; this becomes G_{k+1}.

This process will yield more than 80,000 random numbers G_k before the sequence becomes periodic and the numbers begin to repeat. Various tests on the first 50,000 numbers give completely satisfactory results. These numbers have been used in solving a wide variety of problems.

The advantages of the pseudo-random number method are quite evident. First, to obtain each number requires only a few simple operations, so that the speed of generation of random numbers is on the same order as the computer's work speed. Second, the program occupies very little space in the computer's memory. Third, the sequence of G_k can be easily reproduced. Finally, it is only necessary to verify the "quality" of such a series once; after that, it can be used many times for calculations in suitable problems without fear of error.

The single disadvantage of this method is the limited supply of pseudo-random numbers which it gives. However, there are ways to obtain still more of them. In particular, it is possible to change the initial number G_0.

The overwhelming majority of computations currently performed by the Monte Carlo method use pseudo-random numbers.

4

Transformations of Random Variables

The necessity of simulating different random variables arises in solving various problems. In the early stages of the use of the Monte Carlo method, some experimenters tried to construct a wheel for finding

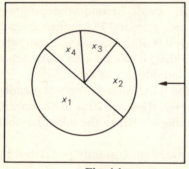

Fig. 4.1

each random variable. For example, in order to find values of a random variable with the distribution

$$\begin{pmatrix} x_1 & x_2 & x_3 & x_4 \\ 0.5 & 0.25 & 0.125 & 0.125 \end{pmatrix}, \quad (4.1)$$

one would use the wheel illustrated in figure 4.1, which operates in the same way as the wheel in figure 3.1, but which has unequal divisions, in the proportions p_i.

However, this turns out to be completely unnecessary. Values for any random variable can be obtained by transformations on the values of one "standard" random variable. Usually this role is played by G, the uniform distribution over the interval [0, 1]. We already know how to get the values of G.

The process of finding the values of some random variable X, by transforming one or more values of G, we will call the *construction* of X.

4.1. Constructing a Discrete Random Variable

Assume that we want to obtain values of a random variable X with the distribution

$$X = \begin{pmatrix} x_1 & x_2 & \cdots & x_n \\ p_1 & p_2 & \cdots & p_n \end{pmatrix}.$$

24

Let us examine the interval $0 \le y \le 1$ and break it up into n intervals with lengths of $p_1, p_2, \ldots, p_n$. The coordinates of the points of division will obviously be $y_1 = p_1$, $y_2 = p_1 + p_2$, $y_3 = p_1 + p_2 + p_3, \ldots,$ $y_{n-1} = p_1 + p_2 + \cdots + p_{n-1}$.

We number the resulting intervals $1, 2, \ldots, n$ (fig. 4.2):

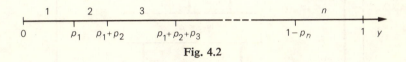

Fig. 4.2

Each time we need to "perform an experiment" and to select a value of X, we shall choose a value of G and find the point $y = G$. If this point lies in the interval numbered i, we will consider that $X = x_i$ (for this trial).

It is easy to demonstrate the validity of such a procedure. Since the random variable G is uniformly distributed over $[0, 1]$, the probability that a G is in any interval is equal to the length of that interval. That is,

$$P(0 \le G < p_1) = p_1 ,$$
$$P(p_1 \le G < p_1 + p_2) = p_2 ,$$
$$\vdots$$
$$P(p_1 + p_2 + \cdots + p_{n-1} \le G \le 1) = p_n .$$

According to our procedure, $X = x_i$ whenever

$$p_1 + p_2 + \cdots + p_{i-1} \le G < p_1 + p_2 + \cdots + p_i ,$$

and the probability of this event is p_i.

Of course, on a computer we can get along without figure 4.2. Let us assume that the numbers $x_1, x_2, \ldots, x_n$ have been placed in successive storage locations in the memory, and likewise the probabilities $p_1, p_1 + p_2, p_1 + p_2 + p_3, \ldots, 1$. A flow chart of the subroutine for the construction of X is provided in figure 4.3.

Example. To construct ten values of a random variable T with the distribution

$$T = \begin{pmatrix} 3 & 4 \\ 0.58 & 0.42 \end{pmatrix} .$$

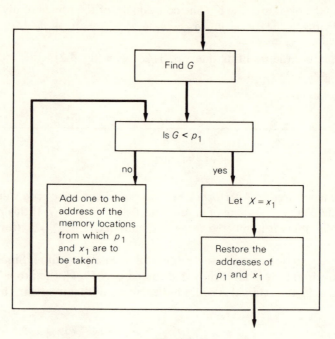

Fig. 4.3

For the values of G we take pairs of numbers from table A in the Appendix multiplied by 0.01.[1] Thus, $G = 0.86$; 0.51; 0.59; 0.07; 0.95; 0.66; 0.15; 0.56; 0.64; 0.34.

Clearly, under our procedure the values of G less than 0.58 correspond to the value $T = 3$, and the values of $G \geq 0.58$ to the value $T = 4$. Thus, we obtain the values $T = 4$; 3; 4; 3; 4; 4; 3; 3; 4; 3.

Note that the order of the values $x_1, x_2, \ldots, x_n$ in the distribution X is arbitrary, although it should be the same throughout the construction.

4.2. The Construction of Continuous Random Variables

Now let us assume that we need to get values of a random variable X which is distributed over the interval $[a, b]$ with density $p(x)$.

---

1. Since in this example the p_i are given to two decimal places, it suffices to take the values of G to two decimal places. In an approximation of this sort, where the case of $G = 0.58$ is possible, it should be included with the case $G > 0.58$ (for the value $G = 0.00$ is possible, but not the value $G = 1.00$). When more decimal places for G are used, the case of the equality $G = p_i$ is improbable, and it can be included in either of the inequalities.

We shall prove that values of X are given by the equation

$$\int_a^X p(x)\,dx = G\,;\qquad\qquad(4.2)$$

that is, taking each value of G in turn, we must solve equation (4.2) and find the corresponding value of X.

For the proof let us examine the function (fig. 4.4)

$$y = \int_a^X p(x)\,dx\,.$$

From the general properties of density (2.15) and (2.16), it follows that

$$y(a) = 0\,,\qquad y(b) = 1\,,$$

and, taking the derivative,

$$y'(x) = p(x) \geq 0\,.$$

This means that the function $y(x)$ increases monotonically from 0 to 1. Furthermore, almost any line $y = G$, where $0 \leq G \leq 1$, intersects the curve $y = y(x)$ in one and only one point, the abscissa of which we take as X. If we agree to take for values of G lying on "flat spots" on the curve, the value of X corresponding to one of the endpoints of the flat spot, then equation (4.2) will always have one and only one solution.

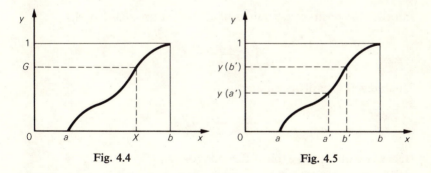

Fig. 4.4   Fig. 4.5

Now we take an arbitrary interval (a', b'), contained in $[a, b]$. The points of this interval

$$a' < x < b'$$

correspond to those ordinates of the curve $y = y(x)$ which satisfy the inequality

$$y(a') < y < y(b'),$$

or to possible "flat spots" with ordinates $y(a')$ and $y(b')$. Since the derivative $y'(x) = p(x)$ is zero everywhere on these "flat spots," they contribute nothing to the probability $P(a' < X < b')$, and therefore (fig. 4.5),

$$P(a' < X < b') = P(y(a') < G < y(b')).$$

Since G is evenly distributed over $(0, 1)$,

$$P(y(a') < G < y(b')) = y(b') - y(a') = \int_{a'}^{b'} p(x)\, dx.$$

Therefore,

$$P(a' < X < b') = \int_{a'}^{b'} p(x)\, dx,$$

and this means exactly that the random variable X, which is a root of equation (4.2), has the probability density $p(x)$.

Example. The random variable H is said to be *uniformly distributed over the interval* $[a, b]$ if its density is constant in this interval:

$$p(x) = \frac{1}{b - a} \quad \text{for all } a < x < b.$$

In order to construct the values of H, we set up equation (4.2):

$$\int_a^H \frac{dx}{b - a} = G.$$

The integral is easily computed:

$$\frac{H - a}{b - a} = G.$$

Hence, we obtain an explicit formula for H:

$$H = a + G(b - a). \tag{4.3}$$

Other examples of the application of formula (4.2) will be given in sections 5.2 and 8.3.

4.3. Neyman's Method for the Construction of Continuous Random Variables

It can prove exceedingly difficult to solve equation (4.2) for X; for example, when the integral of $p(x)$ is not expressed in terms of elementary functions, or when the density of $p(x)$ is given graphically.

Let us suppose that the random variable X is defined over a finite interval (a, b) and its density is bounded (fig. 4.6):

$$p(x) \leq M_0 .$$

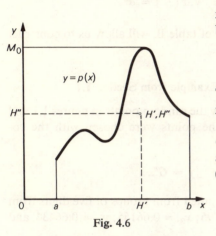

Fig. 4.6

The value of X can be constructed in the following way:

(1) We take two values G' and G'' of the random variable G and locate the random point (H', H'') with coordinates

$$H' = a + G'(b - a),$$
$$H'' = G''M_0 .$$

(2) If this point lies under the curve $y = p(x)$, then we set $x = H'$; if it lies above the curve, we reject the pair (G', G'') and select a new pair of values.

The justification for this method is presented in section 9.1.

4.4. On Constructing Normalized Variables

There are many ways of constructing the various random variables. We shall not deal with all of them here. They are usually not used unless the methods of sections 4.2 and 4.3 prove ineffective.

Specifically, this happens in the case of a normalized variable Z, since the equation

$$\frac{1}{\sqrt{(2\pi)}} \int_{-\infty}^{Z} \exp\left(-\frac{x^2}{2}\right) dx = G$$

is not explicitly solvable, and the interval containing possible values of Z is infinite.

In table B in the Appendix, values, already constructed, are given for a normal random variable Z with mathematical expectation $E(Z) = 0$

and variance $\mathrm{Var}(Z) = 1$. It is not hard to prove[2] that the random variable

$$Z' = a + \sigma Z \qquad (4.4)$$

will also be normal and, moreover, it follows from (10) and (11) that

$$E(Z') = a, \qquad \mathrm{Var}(Z') = \sigma^2.$$

Thus, formula (2.2), with the help of table B, will allow us to construct any normal variable.

4.5. More About the Example from Section 1.1

Now it is possible to explain how the random points in figures 1.1 and 1.2 were selected. In figure 1.1 the points were chosen with the co-ordinates

$$x = G', \qquad y = G''.$$

The values of G' and G'' were computed from groups of five digits from table A: $x_1 = 0.86515$; $y_1 = 0.90795$; $x_2 = 0.66155$; $y_2 = 0.66434$, and so on.

It can be proved[3] that since the abscissas and the ordinates of these points are independent, the probability of hitting a point in any region within the square is equal to the area of the region. Stated differently, this means that the points are uniformly distributed over the square.

In figure 1.2 the points were made with the coordinates

$$x = 0.5 + 0.2Z', \qquad y = 0.5 + 0.2Z'',$$

where the values of Z' and Z'' were taken successively from table B:

$$x_1 = 0.5 + 0.2 \cdot 0.2005, \qquad y_1 = 0.5 + 0.2 \cdot 1.1922;$$

$$x_2 = 0.5 + 0.2(-0.0077), \ldots.$$

One of the points, falling outside the square, was discarded.

From formula (4.4) it follows that the abscissas and ordinates of these points are normal random variables with means $a = 0.5$ and variances $\sigma^2 = 0.04$.

2. Proof is given in section 9.2.
3. Proof is given in section 9.3.

Part 2

Examples of the Application of the Monte Carlo Method

Simulating
a Mass-Supply
System

5.1. Description of the Problem

Let us examine one of the simplest mass-supply systems. Consider a system like the check-out section of a supermarket, consisting of n lines (or channels, or distribution points), each of which can "wait on customers." Demands come into the system, the moments of their entrances being random. Each demand starts on line number 1. If this line is free at time T_k, when the kth demand enters the system, it will begin to supply the demand, a process lasting a time t. If at the instant T_k line 1 is busy, the demand is instantly transferred to line 2, and so on. Finally, if all n lines are busy at the instant T_k, the system is said to overflow.

Our problem is to determine how many demands (on the average) the system satisfies in an interval of time T and how many times it will overflow.

Problems of this type are encountered constantly in the research of market organizations, and not only those providing everyday services. In some very special cases it is possible to find an analytical solution; but in complex situations like those we shall describe later, the Monte Carlo method turns out to be the only possible method of calculation.

5.2. The Simple Demand Flow

The first question which comes up in our examination of this system is: What is the form of the flow of incoming demands? This question is usually answered by observations of the system, or of similar systems, over long periods of time. From the study of demand flows under various conditions we can select some frequently encountered cases.

The *simple*, or *Poisson*, demand flow occurs when the interval of time

S between two consecutive demands is a random variable, distributed over the interval $[0, \infty)$ with density

$$p(x) = ae^{-ax}. \tag{5.1}$$

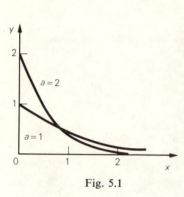

Fig. 5.1

Formula (5.1) is also called the *exponential distribution* (see fig. 5.1, where the densities (5.1) are constructed for $a = 1$ and $a = 2$).

It is easy to compute the mathematical expectation of S:

$$E(S) = \int_0^\infty xp(x)\,dx = \int_0^\infty xae^{-ax}\,dx.$$

After integrating by parts ($u = x$, $dv = ae^{-ax}\,dx$), we obtain

$$E(S) = [-xe^{-ax}]_0^\infty + \int_0^\infty e^{-ax}\,dx = \left[-\frac{e^{-ax}}{a}\right]_0^\infty = \frac{1}{a}.$$

The parameter a is called the *demand flow density*.

The formula for constructing S is easily obtained from equation (4.2), which in the present case is written:

$$\int_0^S ae^{-ax}\,dx = G.$$

Computing the integral on the left, we get the relation

$$1 - e^{-aS} = G,$$

and, hence,

$$S = -\frac{1}{a}\ln(1 - G).$$

The variable $1 - G$ has exactly the same distribution as G, and so, instead of this last formula, one can use the formula

$$S = -\frac{1}{a}\ln G. \tag{5.2}$$

5.3. The Plan for the Computation

Let us look at the operation of the system of section 5.1 in the case of the simple demand flow.

To each line we assign a storage location in the memory of a computer, in which we shall register the moment the line becomes free. Let us designate the next time at which the ith line will become free by t_i. At the beginning of the calculation we let the time when the first demand enters the system, T_1, equal zero. One can see that at this point all the t_i are equal to 0; all the lines are free. The calculation ends at time $T_f = T_1 + T$.

The first demand enters line 1. This means that for the period t this line will be busy. Therefore, we should substitute for t_1 the new value $(t_1)_{new} = T_1 + t$, add one to the counter of demands met, and return to examine the second demand.

Let us assume that k demands have already been examined. It is necessary, then, to select the time for the entrance of the $(k + 1)$th demand. For this we take the next value of G and compute the next value of S (S_k) by formula (5.2). Then we compute the entrance time

$$T_{k+1} = T_k + S_k .$$

Is the first line free at this time? To establish this it is necessary to verify the condition

$$t_1 \leq T_{k+1} . \tag{5.3}$$

If this condition is met, it means that at time T_{k+1} the line is free and can attend to the demand. We therefore replace t_1 by $T_{k+1} + t$, add one to the counter, and return for the next demand.

If condition (5.3) is not met, it means that at T_{k+1} the first line is busy. Then we test whether the second line is free:

$$t_2 \leq T_{k+1} ? \tag{5.4}$$

If condition (5.4) is met, we replace t_2 by $T_{k+1} + t$, add one to the counter, and go on to the next demand.

If condition (5.4) is not met either, we proceed to a test of the condition

$$t_3 \leq T_{k+1} .$$

It can happen that for all i from 1 to n,

$$t_i > T_{k+1} ,$$

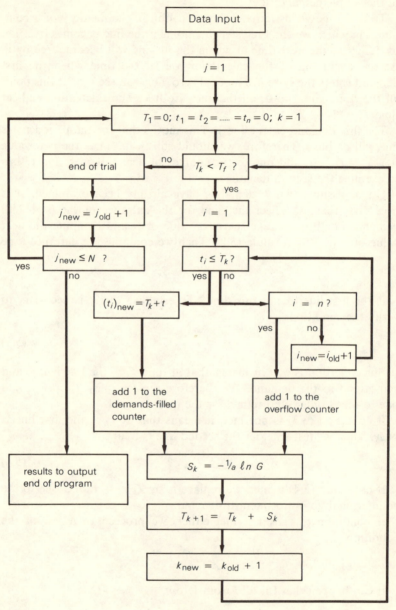

Fig. 5.2

that is, that at time T_{k+1} all the lines are busy. In this case we add one to the overflow counter and go on to examine the next demand.

Each time T_{k+1} is computed, it is necessary to test the condition for the termination of the experiment:

$$T_{k+1} > T_f.$$

When this condition is satisfied, the trial comes to an end. On the counters are the number of demands successfully met (m_d) and the number of overflows (m_0).

Let this experiment be repeated N times. Then the results of all the trials are averaged:

$$E(m_d) \approx \frac{1}{N} \sum_{j=1}^{N} (m_d)_j,$$

$$E(m_0) \approx \frac{1}{N} \sum_{j=1}^{N} (m_0)_j,$$

where $(m_d)_j$ and $(m_0)_j$ are the values of m_d and m_0 obtained on the jth trial.

In figure 5.2 a flow chart of the program which performs this calculation is given. (If the values of m_d and m_0 for single trials are desired, they can be printed out in the square marked "end of trial.")

5.4. More Complex Problems

It is easy to see that we can use this method to compute results for more complex systems. For example, the value t, rather than being fixed, can be different for the various lines (this would correspond to different equipment or to varying qualifications of the service staff), or a random variable whose distribution differs for the various lines. The plan for the calculation remains roughly the same. The only change is that a new value of t is generated for each demand and the formula for each line is independent of that for the others.

One can also examine so-called *waiting-time systems*, which do not overflow immediately. The demand is stored for a short period t' (its *waiting time in the system*), and, if any line becomes available during that time, it attends to that demand.

Systems can also be considered in which the next demand is taken on by the line which will first become available. It is possible to allow for random variations in the density of the demand flow over time, for a random repair time on each line, and many other possibilities.

Of course, such simulations are not done effortlessly. In order to obtain results of any practical value, one must choose a sound model, and this requires extremely careful study of the actual demand flows, time-study observations of the work at the various distribution points, and so on.

In order to study any system of this type, one must know the probabilistic principles of the functioning of the various parts of the system. Then the Monte Carlo method permits the computation of the probabilistic principles of the entire system, however complex it may be.

Such methods of calculation are extremely helpful in planning enterprises. Instead of a costly (and sometimes impossible) real experiment, we can conduct experiments on a computer, trying out different methods of job organization and of equipment usage.

6

Calculating the Quality and Reliability of Products

6.1. The Simplest Plan for Quality Calculation

Let us examine a product S, made up of (perhaps many) elements. For example, S may be a piece of electrical equipment, made of resistors $(R_{(k)})$, capacitors $(C_{(k)})$, tubes, and the like. We define the quality of the product as the value of a single output parameter U, which can be computed from the parameters of all the elements:

$$U = f(R_{(1)}, R_{(2)}, \ldots; C_{(1)}, C_{(2)}, \ldots; \ldots). \qquad (6.1)$$

If, for example, U is the voltage in an operating section of an electric circuit, then by Ohm's law it is possible to construct equations for the circuit and, solving them, to find U.

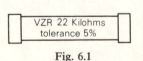

VZR 22 Kilohms
tolerance 5%

Fig. 6.1

In reality the parameters of the elements of a mechanism are never exactly equal to their indicated values. For example, the resistor illustrated in figure 6.1 can test out anywhere between 20.0 and 23.1 kilohms.

The question arises: What effect do deviations of the parameters of all these elements have on the value of U?

One can try to compute the limits of the dimension U, taking the "worst" values of the parameters of each element. However, it is not always clear which values will be the worst. Furthermore, if the number of elements is large, the limits thus computed will be highly over-estimated, for it is unlikely that all the parameters will be simultaneously at their worst.

Therefore, it is more reasonable to calculate the parameters of all the elements and the value of U itself by the Monte Carlo method and to try to estimate its mathematical expectation $E(U)$ and variance

Var (U). $E(U)$ will be the mean value of U for all the parts of the product, and Var (U) will show how much deviation of U from $E(U)$ will be encountered in practice.

Recall (see section 2.2) that, in general,

$$E(U) \neq f(E(R_{(1)}), E(R_{(2)}), \ldots; E(C_{(1)}), E(C_{(2)}), \ldots; \ldots).$$

It is practically impossible to compute analytically the distribution of U for any function f which is at all complex. Sometimes this can be done experimentally by looking at a large lot of finished products. But even this is not always possible, certainly not in the design stage.

Let us try to apply our method. To do so, we shall need to know: (a) the probabilistic characteristics of all the elements, and (b) the function f (more exactly, a way to compute the value of U from any fixed values $R_{(1)}, R_{(2)}, \ldots; C_{(1)}, C_{(2)}, \ldots; \ldots$).

The probability distribution of the parameters of each single element can be obtained experimentally by examining a large lot of such elements. Quite often the distribution is found to be normal. Therefore, many experimenters proceed in the following way. They consider the resistance of the element pictured in figure 6.1 to be a normal random variable Q with mathematical expectation $E(Q) = 22$ and with $3\sigma = 1.1$ (remember that, according to (2.20), it is rare to get a value of Q deviating from $E(Q)$ by more than 3σ on any one trial).

The plan for the calculation is quite simple. For each element a value of its parameter is constructed; then the value of U is computed according to formula (6.1). Repeating the trial N times and obtaining values $U_1, U_2, \ldots, U_N$, we can compute that, approximately,

$$E(U) \approx \frac{1}{N} \sum_{j=1}^{N} U_j,$$

$$\mathrm{Var}\,(U) \approx \frac{1}{N-1} \left[\sum_{j=1}^{N} (U_j)^2 - \frac{1}{N} \left(\sum_{j=1}^{N} U_j \right)^2 \right].$$

For large N in the latter formula one can replace the factor $1/(N-1)$ by $1/N$, and then this formula is a simple consequence of formulas (2.8) and (2.9). In statistics it has been shown that for small N it is better to keep the factor $1/(N-1)$.

6.2. Examples of the Calculation of Reliability

Suppose we want to estimate how long, on the average, a product will function properly, assuming that we know the relevant characteristics of each of its components.

If we consider the breakdown time of each component $t_{(k)}$ to be a constant, then computing the breakdown time of the product presents no difficulties. For example, for the product schematically represented in figure 6.2, in which the breakdown of one component implies the breakdown of the entire product,

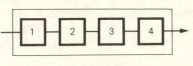

Fig. 6.2

$$t = \min (t_{(1)}; t_{(2)}; t_{(3)}; t_{(4)}). \quad (6.2)$$

And for a product, schematically represented in figure 6.3, where one of the elements is duplicated, or *redundant*,

$$t = \min [t_{(1)}; t_{(2)}; \max (t_{(3)}, t_{(4)}); t_{(5)}], \quad (6.3)$$

since if element 3 fails, for example, the product will continue to work with the single element 4.

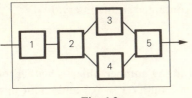

Fig. 6.3

In actual practice the breakdown time of any component k of a mechanism takes the form of a random variable $F_{(k)}$. When we say that a light bulb is good for 1,000 hours, we only mean that this is the average value $E(F)$ of the variable F. Everyone knows that one bulb may burn out sooner than another one like it.

If the density distribution $F_{(k)}$ is known for each of the components of the product, $E(F)$ can be computed by the Monte Carlo method, following the plan of section 6.1. That is, for each element it is possible to construct a value of the variable F_k; let us call it f_k. Then it is possible to compute a value f of the random variable F, representing the breakdown time of the entire product, by a formula corresponding to (6.2) or (6.3). Repeating this experiment enough times (N), we can obtain the approximation

$$E(F) \approx \frac{1}{N} \sum_{j=1}^{N} f_j,$$

where f_j is the value f obtained on the jth trial.

It must be noted that the question of the distributions $F_{(k)}$ of breakdown times for the various elements is not at all a simple one. For long-lived elements, actual experiments to determine the distributions are difficult to perform, since one must wait until enough of the elements have broken down.

6.3. Further Possibilities of the Method

The preceding examples show that the procedure for calculating the quality of products being designed is quite simple in theory. We must know the probabilistic characteristics of all the components of the product, and we must succeed in computing the variable in which we are interested as a function of the parameters of these components. Then we can allow for the randomness of the parameters by means of our simulation.

From the simulation it is possible to obtain much more useful information than just the mean and the variance of the variable that interests us. Suppose that we have obtained a large number of values $U_1, U_2, \ldots,$ U_N of the random variable U. From these values we can construct the approximate density distribution of U. In the most general cases, this is a rather difficult statistical question. Let us limit ourselves, then, to a concrete example.

Suppose that we have, all together, 120 values $U_1, U_2, \ldots, U_{120}$ of the random variable U, all of them contained in the interval

$$1 < U_j < 6.5 .$$

We break this interval into eleven (or any number which is neither too large nor too small) equal intervals of length $\Delta x = 0.5$ and count how many values of U_j fall in each interval. The results are given in figure 6.4.

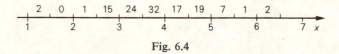

Fig. 6.4

The frequency of hits in any interval yields the proportion of hits in that interval out of $N = 120$. In our example the frequencies are: 0.017; 0; 0.008; 0.12; 0.20; 0.27; 0.14; 0.16; 0.06; 0.008; 0.017.

On each of the intervals of the partition, let us construct a rectangle with area equal to the frequency of values of U_j falling in that interval (fig. 6.5). In other words, the height of each rectangle will be equal to the frequency divided by Δx. The resulting graph is called a *histogram*.

The histogram serves as an approximation to the unknown density of the random variable U. Therefore, for example, the area of the histogram bounded by $x = 2.5$ and $x = 5.5$ gives us an approximate value for the probability

$$P(2.5 < U < 5.5) \approx 0.95 .$$

On the basis of the above calculation (the trial), it is possible to estimate that there is a probability of 0.95 that a value of U will fall in the interval $2.5 < U < 5.5$.

In figure 6.5 the density of a normal random variable Z' with the parameters $a = 3.85$, $\sigma = 0.88$ has been constructed as a comparison.[1]

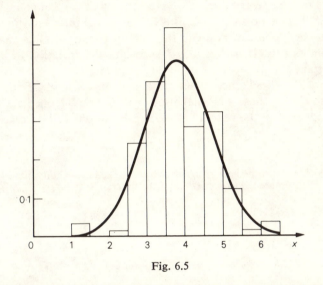

Fig. 6.5

If we now compute the probability that Z' falls within the interval $2.5 < Z' < 5.5$ for this density, we get the "fitted" value 0.91.[2]

---

1. The numbers $a = 3.85$ and $\sigma = 0.88$ were obtained by considering a random variable with the distribution

$$\begin{pmatrix} x_1 & x_2 & x_3 & \cdots & x_{11} \\ 0.017 & 0 & 0.008 & \cdots & 0.017 \end{pmatrix}, \qquad (*)$$

where each x_k is the value of the midpoint of the kth interval (thus, $x_1 = 1.25$, $x_2 = 1.75$, and so on), and then calculating the expectation a and the variance σ^2 of such a random variable by formulas (2.3) and (2.9). This process is called *fitting* a normal density to the frequency distribution $(*)$.

2. Here is the method used to compute this value. In accordance with (2.14), we write

$$P(2.5) < Z' < 5.5) = \frac{1}{\sigma\sqrt{(2\pi)}} \int_{2.5}^{5.5} \exp\left[-\frac{(x-a)^2}{2\sigma^2}\right] dx .$$

In the integral we make a substitution for the variable $(x-a)/\sigma = t$. Then we obtain

$$P(2.5 < Z' < 5.5) = \frac{1}{\sqrt{(2\pi)}} \int_{t_1}^{t_2} \exp\left(-\frac{t^2}{2}\right) dt ,$$

where $t_1 = (2.5 - a)/\sigma = -1.54$ and $t_2 = (5.5 - a)/\sigma = 1.88$. The latter integral

6.4. A Remark

It is unfortunate that calculations of this type are not at present performed more commonly. It is difficult to say why this is so. Most likely it is because designers and planners are not aware of the possibility.

Moreover, before using the method to simulate any product, one must find out the probabilistic characteristics of all the components that go into it. This is no small task. But it is also true that, knowing these characteristics, one can evaluate the quality of *any* product made of these components. It is even possible to find the variation in quality when certain components are replaced by others.

The probabilistic characteristics of the elements will always be a prominent obstacle for those who make such calculations. Nonetheless, one might hope that in the near future such calculations will become more usual.

---

can be evaluated with the help of tables of the so-called probability integral $\phi(x)$, in which are given the values for $x \geq 0$ of the function

$$\phi(x) = \frac{1}{\sqrt{(2\pi)}} \int_{-\infty}^{x} \exp\left(-\frac{t^2}{2}\right) dt .$$

We obtain

$$P(2.5 < Z' < 5.5) = \phi(1.54) + \phi(1.88) - 1 = 0.91,$$

using the identity $\phi(x) + \phi(-x) = 1$, which can easily be verified by looking at the graph of the normal distribution

$$P(x) = \frac{1}{\sqrt{(2\pi)}} \exp\left(-\frac{x^2}{2}\right).$$

7

Simulating the Penetration of Neutrons through a Block

The laws of probability, as they apply to interactions of single elementary particles (neutrons, photons, mesons, and others) with matter, are known. Usually it is necessary to find out the macroscopic characteristics of these processes, those in which an enormous number of such particles participate: density, current flow, and so on. This situation is similar to the one we met in chapters 5 and 6, and it, too, can be handled by the use of the Monte Carlo method.

Most frequently, perhaps, the Monte Carlo method is used in the study of the physics of neutrons. We shall examine an elementary variant of the problem of the penetration of neutrons through a block.

7.1. A Formulation of the Problem

Let a stream of neutrons with energy E_0 fall at an angle of 90° on a homogeneous block of infinite extent but of finite depth h. In collisions with atoms of the matter of which the block is composed, neutrons can be deflected elastically or absorbed. Let us assume, for simplicity, that the energy of a neutron does not change when it is deflected, and that a neutron will "rebound" off an atom in any direction with equal probability. This is approximately the case for matter composed of heavy atoms. The histories of several neutrons are portrayed in figure 7.1: neutron (a) penetrated the block, neutron (b) is absorbed, neutron (c) is reflected from the block.

We are required to compute the probability p^+ of a neutron penetrating the block, the probability p^- of a neutron being reflected from the block, and the probability p^0 of a neutron being absorbed by the block.

45

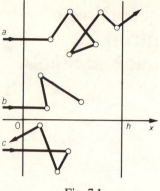

Fig. 7.1

Interaction of neutrons with matter is characterized in the case under consideration by two constants $\sum_c$ and $\sum_s$, respectively, called the *absorption cross-section* and the *dispersion cross-section*. The subscripts c and s are the initial letters of the words "capture" and "scattering."

The sum of these cross-sections is called the *total cross-section*

$$\sum = \sum_c + \sum_s .$$

The physical significance of the cross-sections is this: In a collision of a neutron with an atom of matter the probability of absorption is equal to $\sum_c/\sum$, and the probability of reflection is $\sum_s/\sum$.

The *free path length L* of a neutron (that is, the distance between consecutive collisions) is a random variable. We shall assume that it can take any positive value from a probability density

$$p(x) = \sum e^{-\sum x} .$$

This density of the variable L coincides with the density (5.1) of the random variable S for the simple demand flow. By analogy with section 5.2 we can immediately write the expression for the mean free-path length

$$E(L) = 1/\sum$$

and the formula for constructing L:

$$L = -(1/\sum) \ln G .$$

There remains to be clarified the question of how to select the random direction of the neutron after the collision. Since the situation is symmetric with respect to the x-axis, the direction can be defined as the single angle ϕ formed by the final direction of the velocity of the neutron and the x-axis. It can be proved [1] that the necessity of having equal probabilities in each direction is in this case equivalent to its being necessary that the cosine of this angle, $M = \cos \phi$, be uniformly distributed over

1. Proof is given in section 9.4.

the interval $[-1, 1]$. From formula (4.3), letting $a = -1$, $b = 1$, the formula for constructing M follows:

$$M = 2G - 1.$$

7.2. A Plan for the Calculation by Means of the Simulation of Real Trajectories

Let us assume that a neutron underwent its kth deflection inside the block at the point x_k and afterwards began to move in the direction M_k.

Let us construct the free-path length

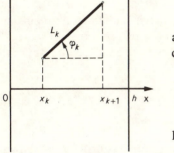

$$L_k = -(1/\Sigma) \ln G$$

and compute the abscissa of the next collision (fig. 7.2)

$$x_{k+1} = x_k + L_k M_k.$$

We check to see if the condition for penetrating the block has been met:

Fig. 7.2

$$x_{k+1} > h.$$

If it has, the calculation of the neutron's trajectory stops, and a 1 is added to the counter for penetrated particles. Otherwise, we test the condition for reflection:

$$x_{k+1} < 0.$$

If this condition is met, the calculation of the neutron's trajectory stops and a 1 is added to the counter for reflected particles. If this condition also fails, that is, if $0 \le x_{k+1} \le h$, it means that the neutron has undergone its $(k + 1)$th collision within the block, and it is necessary to construct the effect of this collision on the neutron.

In accordance with the method of section 4.1, we take the next value of G and test the condition for absorption:

$$G < \Sigma_c/\Sigma.$$

If this last inequality holds, then the calculation of the neutron's trajectory stops, and a 1 is added to the counter for absorbed particles. If not, we consider that the neutron has undergone a deflection at the point x_{k+1}. Then we generate a new direction of movement

$$M_{k+1} = 2G - 1$$

and repeat the cycle once more (using different values of G, of course).

All the G are written without subscripts, since each value of G is used only once. Up to three values of G are needed to calculate each jog of the trajectory.

The initial values for every trajectory are:

$$x_0 = 0, \qquad M_0 = 1.$$

After N trajectories have been computed, it is found that N^+ neutrons have gone through the block, N^- have been reflected from it, and N^0 have been absorbed. Obviously, the desired probabilities are approximately equal to the ratios

$$p^+ \approx \frac{N^+}{N}, \qquad p^- \approx \frac{N^-}{N}, \qquad p_0 \approx \frac{N^0}{N}.$$

In figure 7.3 a flow chart of the program for this problem is shown. The subscript j is the number of the trajectory, and the subscript k is the collision number along the trajectory.

This computation procedure, although it is very natural, is not perfect. In particular, it is difficult to determine the probabilities p^+ and p^- by this method when they are very small. This is precisely the case one encounters in calculating protection against radiation.

However, by more sophisticated applications of the Monte Carlo method, even these computations are possible. We will briefly consider one of the simplest variants of calculation with the help of so-called "weights."

7.3. A Plan for the Calculation Using Weights to Avoid Terminal Absorption

Let us reexamine the problem of neutron penetration. Let us assume that a "package," consisting of a large number w_0 of individual neutrons, is traveling along a single trajectory. For a collision at the point x_1 the average number of neutrons in the package which would be absorbed is $w_0 \Sigma_c / \Sigma$, and the number of neutrons undergoing deflection would be, on the average, $w_0 \Sigma_s / \Sigma$.

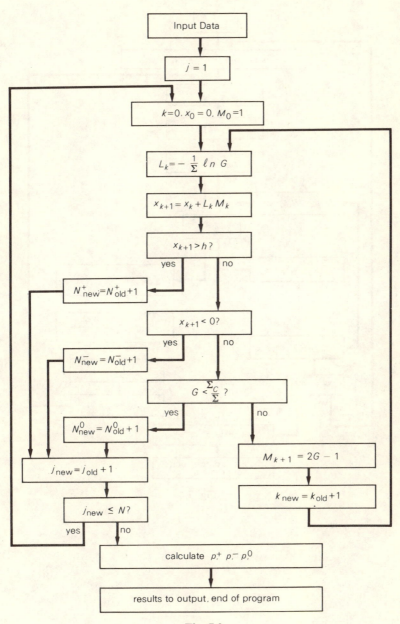

Fig. 7.3

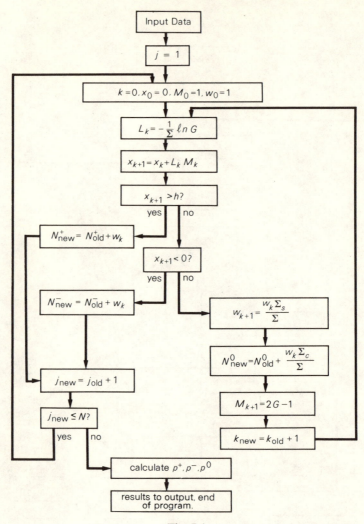

Fig. 7.4

In our program, after each collision, we therefore add the value $w_0 \Sigma_c / \Sigma$ to the absorbed-particle counter, and watch the motion of the deflected package, assuming that the entire remainder of the package is deflected in a single direction.

All the formulas for the calculation given in section 7.2 remain the same. For each collision the number of neutrons in the package is simply reduced:

$$w_{k+1} = \frac{w_k \Sigma_s}{\Sigma},$$

since that part of the package comprising $w_k \Sigma_c / \Sigma$ neutrons will be absorbed. Now the trajectory cannot be ended by absorption.

The value w_k is usually called the *weight* of the neutron and, instead of talking about a "package" consisting of w_k neutrons, one speaks of a neutron with weight w_k. The initial weight w_0 is usually set equal to 1. This does not conflict with our notion of a "large package," since all the w_k obtained while computing a trajectory contain w_0 as a common factor.

A flow chart of the program which realizes this calculation is given in figure 7.4. It is no more complex than the flow chart in figure 7.3. It is possible to prove,[2] however, that calculating p^+ by this method is always more efficient than using the method of section 7.2.

7.4. A Remark

There are a great many other ways to do the calculation, using various weights, but we cannot stop to consider them here. We simply stress that the Monte Carlo method enables one to solve many complex problems about elementary particles. The medium used can consist of any substance and can have any geometrical structure; the energy of the particles can, if we so desire, be changed with each collision. It is possible by this technique to simulate many other nuclear processes. For example, we can construct a model for the fissioning of an atom and the formation of new neutrons by collision with a neutron, and thus simulate the conditions for the initiation and maintenance of a chain reaction. Problems related to this were, in fact, among the first serious applications of the Monte Carlo method to scientific problems.

---

2. Proof is given in section 9.5.

8

Evaluating a Definite Integral

The problems examined in chapters 5, 6, and 7 were probabilistic by nature, and to use the Monte Carlo method to solve them seemed quite natural. Here a purely mathematical problem is considered: the approximate evaluation of a definite integral.

Since evaluating a definite integral is equivalent to finding an area, we could use the method of section 1.2. In this chapter, however, we shall present a more effective method, which allows us to construct several probabilistic models for solving the problem by the Monte Carlo method. We shall finally indicate how to choose the best from among all these models.

8.1. The Method of Computation

Let us examine a function $g(x)$, defined on the interval $a \leq x \leq b$. Our assignment is to compute approximately the integral

$$I = \int_a^b g(x) \, dx . \tag{8.1}$$

We select an arbitrary density distribution $p_V(x)$, also defined on the interval $[a, b]$ (that is, a function $p_V(x)$, satisfying conditions (2.15) and (2.16)).

Finally, besides the random variable V, defined on the interval $[a, b]$ with density $p_V(x)$, we need a random variable

$$H = \frac{g(V)}{p_V(V)} .$$

By (2.18),

$$E(H) = \int_a^b \left(\frac{g(x)}{p_V(x)} \right) p_V(x) \, dx = I .$$

52

Now let us look at N identical random variables $H_1, H_2, \ldots, H_N$, and apply the central limit theorem of section 2.4 to their sum. In this case formula (2.21) is written

$$P\left(\left|\frac{1}{N}\sum_{j=1}^{N} H_j - I\right| < 3\sqrt{\left(\frac{\text{Var}(H)}{N}\right)}\right) \approx 0.997. \qquad (8.2)$$

This last relation means that if we choose N values $V_1, V_2, \ldots, V_N$, then for sufficiently large N,

$$\frac{1}{N}\sum_{j=1}^{N} \frac{g(V_j)}{p_V(V_j)} \approx I. \qquad (8.3)$$

It also shows that there is a very large probability that the error of approximation in (8.3) will not exceed $3\sqrt{(\text{Var}(H)/N)}$.

8.2. How to Choose a Plan for the Calculation

We saw that to compute the integral (8.1), we could use any random variable V, defined over the interval $[a, b]$. In any case

$$E(H) = E\left(\frac{g(V)}{p_V(V)}\right) = I.$$

However, the variance and, hence, the estimate of the error of formula (8.3) are dependent on what variable V we use. That is,

$$\text{Var}(H) = E(H^2) - I^2 = \int_a^b \left(\frac{g^2(x)}{p_V(x)}\right) dx - I^2.$$

It can be shown[1] that this expression is minimized when $p_V(x)$ is proportional to $|g(x)|$.

Of course, we certainly do not want to choose very complex $p_V(x)$, since the procedure for constructing values of V then becomes very laborious. But it is possible to use $g(x)$ as a guide in choosing $p_V(x)$ (for an example, see section 8.3).

In practice integrals of the form (8.1) are not computed by the Monte Carlo method; the quadrature formulas provide a more precise technique. In the transition to multivalued integrals the situation changes. The quadrature formulas become very complex, while the Monte Carlo method remains practically unchanged.

1. Proof is given in section 9.6.

8.3. A Numerical Example

Let us approximately compute the integral

$$I = \int_0^{\pi/2} \sin x \, dx \, .$$

The exact value of this integral is known:

$$\int_0^{\pi/2} \sin x \, dx = [-\cos x]_0^{\pi/2} = 1 \, .$$

We shall use two different random variables V for the calculation: One with constant density $2/\pi$ (that is, a uniform distribution over the interval $[0, \pi/2]$), and one with linear density $p_V(x) = 8x/\pi^2$. Both these densities, together with the function being integrated, are shown in figure 8.1. It is evident that the linear density most closely fulfills the

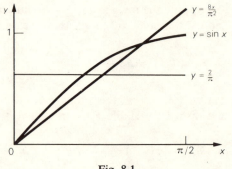

Fig. 8.1

recommendation in section 8.2, that it is desirable for $p_V(x)$ to be proportional to $\sin x$. Therefore, one may expect that it will yield the better result.

(a) Let $p_V(x) = 2/\pi$ on the interval $[0, \pi/2]$. The formula for constructing V can be obtained from formula (4.3) for $a = 0$ and $b = \pi/2$:

$$V = \frac{\pi G}{2} \, .$$

Now formula (8.3) takes the form

$$I \approx \frac{\pi}{2N} \sum_{j=1}^{N} \sin V_j \, .$$

Let $N = 10$. As values of G let us use groups of three digits from table A (multiplied by 0.001). The intermediate results are collected in table 8.1.

Table 8.1

j	1	2	3	4	5	6	7	8	9	10
G_j	0.865	0.159	0.079	0.566	0.155	0.664	0.345	0.655	0.812	0.332
V_j	1.359	0.250	0.124	0.889	0.243	1.043	0.542	1.029	1.275	0.521
$\sin V_j$	0.978	0.247	0.124	0.776	0.241	0.864	0.516	0.857	0.957	0.498

The final result of the computation is:

$$I \approx 0.952 \,.$$

(b) Now let $p_V(x) = 8x/\pi^2$. For the construction of V let us use equation (4.2),

$$\int_0^V \left(\frac{8x}{\pi^2}\right) dx = G \,.$$

After some simple calculations, we obtain

$$V = \frac{\pi}{2} \sqrt{G} \,.$$

Formula (8.3) takes on the form:

$$I \approx \frac{\pi^2}{8N} \sum_{j=1}^{N} \frac{\sin V_j}{V_j} \,.$$

Let $N = 10$. We take the same numbers for G as in (a). The intermediate results are collected in table 8.2.

Table 8.2

j	1	2	3	4	5	6	7	8	9	10
G_j	0.865	0.159	0.079	0.566	0.155	0.664	0.345	0.655	0.812	0.332
V_j	1.461	0.626	0.442	1.182	0.618	1.280	0.923	1.271	1.415	0.905
$\dfrac{\sin V_j}{V_j}$	0.680	0.936	0.968	0.783	0.937	0.748	0.863	0.751	0.698	0.868

The result of the calculation is:

$$I \approx 1.016 \,.$$

As we anticipated, the second method gave the more accurate result.

8.4. On Estimating Error

In section 8.1 it was noted that the absolute value of the error in calculating an integral I practically cannot exceed the value $3\sqrt{(\text{Var}(H)/N)}$. In reality, however, the error as a rule turns out to be noticeably less than this value. Therefore, as a characteristic of error another value is often used in practice—the *probable error*

$$\delta_p = 0.675\sqrt{\left(\frac{\text{Var}(H)}{N}\right)}.$$

Table 8.3

Method	Var (H)	δ_p	δ_c
(a)	0.256	0.103	0.048
(b)	0.016	0.027	0.016

The actual absolute error depends on the particular random numbers used in the calculation and can prove to be twice or three times as large as δ_p, or several times smaller. δ_p gives us, not the upper limit of the error, but rather its order of magnitude. In fact, δ_p is very nearly the value for which a deviation larger than δ_p and a deviation smaller than δ_p are equally likely. To see this, note that we are approximating I by

$$R = \frac{1}{N}\sum H_j.$$

By the central limit theorem of section 2.4, R is approximately a normal random variable with mathematical expectation I and standard deviation $\sigma = \sqrt{(\text{Var}(H)/N)}$. But for any normal random variable Z, it is not hard to calculate that whatever a and σ may be,

$$\int_{a-0.675\sigma}^{a+0.675\sigma} P_z(x)\,dx = 0.5,$$

whence

$$P(|z-a| < 0.675\sigma) = 0.5 = P(|z-a| > 0.675\sigma),$$

that is, deviations from the expected value larger and smaller than the probable error 0.675σ are equally probable.

Let us return to the example in section 8.3. From the values given in tables 8.1 and 8.2, one can approximate the variance Var (H) for both methods of computation. The suitable equation for the calculation was given in section 6.1.[2]

The approximate values of the variance Var (H), the probable errors calculated from them, and the true absolute errors obtained from calculation (δ_c) are shown in table 8.3 for both methods of calculation. We see that δ_c really is on the same order as δ_p.

2. For method (a):

$$\text{Var}(H) = \frac{\pi^2}{9 \cdot 4} \left[\sum_{j=1}^{10} (\sin V_j)^2 - \frac{1}{10} \left(\sum_{j=1}^{10} \sin V_j \right)^2 \right]$$

$$= \frac{\pi^2}{36} (4.604 - 3.676) = 0.256 .$$

For method (b):

$$\text{Var}(H) = \frac{\pi^4}{9 \cdot 64} \left[\sum_{j=1}^{10} \left(\frac{\sin V_j}{V_j} \right)^2 - \frac{1}{10} \left(\sum_{j=1}^{10} \frac{\sin V_j}{V_j} \right)^2 \right]$$

$$= \frac{\pi^4}{576} (6.875 - 6.777) = 0.016 .$$

9

Proofs
of Certain
Propositions

In this chapter demonstrations are given for some assertions made in the preceding chapters. We have gathered them together because they seemed to us somewhat cumbersome for a popular presentation or presupposed knowledge of probability theory.

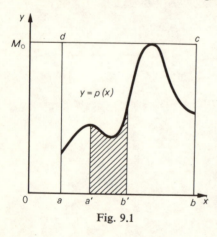

Fig. 9.1

9.1. The Justification of Neyman's Method of Constructing a Random Variable (Section 4.3)

The random point (H', H'') is uniformly distributed over the rectangle $abcd$ (fig. 9.1), the area of which is equal to $M_0(b - a)$.[1] The probability that point (H', H'') is under the curve $y = p(x)$ and will not be discarded is equal to the ratio of the areas

$$\frac{\int_a^b p(x)\,dx}{M_0(b - a)} = \frac{1}{M_0(b - a)}.$$

But the probability that the point is under the curve $y = p(x)$ in the interval $a' < x < b'$ is similarly equal to the ratio of the areas

$$\frac{\int_{a'}^{b'} p(x)\,dx}{M_0(b - a)}.$$

1. Compare section 9.3.

Consequently, among all the values of X that are not discarded, the proportion of values which fall in the interval (a', b') is equal to the quotient

$$\frac{\dfrac{\int_{a'}^{b'} p(x)\, dx}{M_0(b - a)}}{\dfrac{1}{M_0(b - a)}} = \int_{a'}^{b'} p(x)\, dx ,$$

which is what we wanted to show.

9.2. The Density Distribution of a Variable $Z' = a + \sigma Z$
(Section 4.4)

It is assumed that the variable Z is normal, with mathematical expectation $E(Z) = 0$ and variance $\text{Var}(Z) = 1$, so that its density is

$$P_Z(x) = \frac{1}{\sqrt{2\pi}}\, e^{-(x^2/2)} .$$

In order to compute the density distribution of the variable Z', let us choose two arbitrary numbers $x_1 < x_2$ and compute the probability

$$P(x_1 < Z' < x_2) = P(x_1 < a + \sigma Z < x_2)$$

$$= P\left(\frac{x_1 - a}{\sigma} < Z < \frac{x_2 - a}{\sigma}\right) .$$

Consequently,

$$P(x_1 < Z' < x_2) = \frac{1}{\sqrt{2\pi}} \int_{(x_1 - a)/\sigma}^{(x_2 - a)/\sigma} e^{-(x^2/2)}\, dx .$$

We simplify this last integral by substituting the variable $x' = a + \sigma x$. We get

$$P(x_1 < Z' < x_2) = \frac{1}{\sigma\sqrt{2\pi}} \int_{x_1}^{x_2} \exp\left[-(x' - a)^2/2\sigma^2\right] dx' ,$$

whence follows (compare (2.14)) the normality of the variable Z' with parameters $E(Z') = a$, $\text{Var}(Z') = \sigma^2$.

9.3. Uniform Distribution of Points in a Square (Section 4.5)

Since the coordinates of the point (G', G'') are independent, the density $p(x, y)$ is equal to the product of the densities

$$p(x, y) = p_{G'}(x)p_{G''}(y) .^2$$

Each of these densities is identically equal to 1. This means that $p(x, y) = 1$ (for $0 \leq x \leq 1$ and $0 \leq y \leq 1$) and, consequently, the uniformity of distribution of the point (G', G'') in the unit square.

9.4. The Choice of a Random Direction (Section 7.1)

Let us agree to specify a direction by means of a unit vector starting at the origin. The heads of such vectors form the surface of the unit

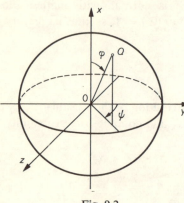

sphere. Now, the words "any direction is equally probable" mean that the head of a vector is a random point Q, uniformly distributed over the surface of the sphere. It follows that the probability of Q lying in any part of the surface dS is equal to $dS/4\pi$.

Let us choose on the surface of the sphere spherical coordinates (ϕ, ψ) (fig. 9.2). Then

$$dS = \sin \phi \, d\phi \, d\psi , \quad (9.1)$$

Fig. 9.2

where $0 \leq \phi \leq \pi$, $0 \leq \psi < 2\pi$.

Since the coordinates ϕ and ψ are independent, the density of the point (ϕ, ψ) is equal to the product $p(\phi, \psi) = p_\phi(\phi)p_\psi(\psi)$. From this equation, relation (9.1), and the relation

$$p(\phi, \psi) \, d\phi \, d\psi = \frac{dS}{4\pi} ,$$

it follows that

$$p_\phi(\phi)p_\psi(\psi) = \frac{\sin \phi}{4\pi} . \quad (9.2)$$

2. This is, in fact, the formal definition of the independence for random variables G' and G''.

Let us integrate this expression with respect to ψ from 0 to 2π. Taking into account the normalizing condition

$$\int_0^{2\pi} p_\psi(\psi) \, d\psi = 1 \,,$$

we obtain

$$p_\phi(\phi) = \frac{\sin \phi}{2} \,. \tag{9.3}$$

Dividing (9.2) by (9.3), we find that

$$p_\psi(\psi) = \frac{1}{2\pi} \,. \tag{9.4}$$

Obviously, ψ is uniformly distributed over the interval $[0, 2\pi)$, and the formula for the construction of ψ will be written thus:

$$\psi = 2\pi G \,. \tag{9.5}$$

We find the formula for the construction of ϕ with the help of equation (4.2):

$$\frac{1}{2} \int_0^\phi \sin x \, dx = G \,,$$

whence

$$\cos \phi = 1 - 2G. \tag{9.6}$$

Formulas (9.5) and (9.6) allow one to select (to construct) a random direction. The values of G in these formulas should, of course, be different.

Formula (9.6) differs from the last formula of section 7.1 only in that G appears in it rather than $1 - G$, but these variables have identical distributions.

9.5. The Superiority of the Method of Weighting (Section 7.3)

Let us introduce the random variables N and N', equal to the number (weight) of neutrons which passed through the block, and obtained by calculating one trajectory by the method of section 7.2 and one by the method of 7.3, respectively.

We know that

$$E(N) = E(N') = p^+.$$

Since N can take on only two values, 0 and 1, the distribution of N is given by the table

$$N = \begin{pmatrix} 1 & 0 \\ p^+ & 1 - p^+ \end{pmatrix}.$$

Taking into account that $N^2 = N$, it is not hard to calculate that Var $(N) = p^+ - (p^+)^2$.

It is easy to see that the variable N' can take on an infinite number of values: $w_0 = 1$, $w_1 = w_0 \Sigma_s / \Sigma$, $w_2 = w_0 (\Sigma_s / \Sigma)^2$, $w_3, \ldots, w_k, \ldots$ and also the value 0 (if the package is reflected from the block instead of passing through). Therefore, its distribution is given by the table

$$N' = \begin{pmatrix} w_0 & w_1 & w_2 & \cdots & w_k & \cdots & 0 \\ q_0 & q_1 & q_2 & \cdots & q_k & \cdots & q \end{pmatrix}.$$

The values q_i need not interest us, since in any case one can write the formula for the variance

$$\text{Var} (N') = \sum_{k=0}^{\infty} w_k^2 q_k - (p^+)^2.$$

Noticing that all the $w_k \le 1$ and that $\sum_{k=0}^{\infty} w_k q_k = E(N') = p^+$, we get the inequality Var $(N') \le p^+ - (p^+)^2 = \text{Var} (N)$.

This fact, that the variance of N' is always less than the variance of N, shows that the method of section 7.3 is always better for calculating p^+ than the method of section 7.2.

The same argument applies to the calculation of p^-, and, if the absorption is not too great, to the calculation of p^0 also.

9.6. The Best Choice for V (Section 8.2)

In section 8.2 we obtained an expression for the variance Var (H). In order to find the minimum of this expression for all possible choices of $p_V(x)$, we make use of an inequality well-known in analysis:

$$\left[\int_a^b |u(x)v(x)| \, dx \right]^2 \le \int_a^b u^2(x) \, dx \cdot \int_a^b v^2(x) \, dx .$$

We set $u = g(x)/\sqrt{(p_V(x))}$ and $v = \sqrt{p_V(x)}$; then from this inequality we obtain

$$\left[\int_a^b |g(x)|\, dx\right]^2 \leq \int_a^b \frac{g^2(x)}{p_V(x)}\, dx \cdot \int_a^b p_V(x)\, dx = \int_a^b \frac{g^2(x)}{p_V(x)}\, dx\, .$$

Thus,

$$\text{Var}\,(H) \geq \left[\int_a^b |g(x)|\, dx\right]^2 - I^2\, . \tag{9.7}$$

It remains to be shown that the lower bound is reached when $p_V(x)$ is proportional to $|g(x)|$.

Let

$$p_V(x) = \frac{|g(x)|}{\int_a^b |g(x)|\, dx}\, . \tag{9.8}$$

It is not hard to compute that for the density $p_V(x)$,

$$\int_a^b \left[\frac{g^2(x)}{p_V(x)}\right] dx = \left[\int_a^b |g(x)|\, dx\right]^2,$$

and the variance $\text{Var}\,(H)$ is really equal to the right side of (9.7).

Let us observe that to take the "best" density (9.8) in the calculation is, in practice, impossible. To get it, it is necessary to know the value of the integral $\int_a^b |g(x)|\, dx$. But the evaluation of this last integral presents a problem just as difficult as the one we are trying to solve: the evaluation of the integral $\int_a^b g(x)dx$. Therefore, we restricted ourselves to the recommendation stated in section 8.2.

Tables

Table A. 400 Random Digits[1]

86515	90795	66155	66434	56558	12332	94377	57802
69186	03393	42502	99224	88955	53758	91641	18867
41686	42163	85181	38967	33181	72664	53807	00607
86522	47171	88059	89342	67248	09082	12311	90316
72587	93000	89688	78416	27589	99528	14480	50961
52452	42499	33346	83935	79130	90410	45420	77757
76773	97526	27256	66447	25731	37525	16287	66181
04825	82134	80317	75120	45904	75601	70492	10274
87113	84778	45863	24520	19976	04925	07824	76044
84754	57616	38132	64294	15218	49286	89571	42903

Table B. 88 Normal Values[2]

0.2005	1.1922	−0.0077	0.0348	1.0423	−1.8149	1.1803	0.0033
1.1609	−0.6690	−1.5893	0.5816	1.8818	0.7390	−0.2736	1.0828
0.5864	−0.9245	0.0904	1.5068	−1.1147	0.2776	0.1012	−1.3566
0.1425	−0.2863	1.2809	0.4043	0.6379	−0.4428	−2.3006	−0.6446
0.9516	−1.7708	2.8854	0.4686	1.4664	1.6852	−0.9690	−0.0831
−0.5863	0.8574	−0.5557	0.8115	−0.2676	−1.2496	−1.2125	1.3846
1.1572	0.9990	−0.1032	0.5405	−0.6022	0.0093	0.2119	−1.4647
−0.4428	−0.5564	−0.5098	−1.1929	−0.0572	−0.5061	−0.1557	−1.2384
−0.3924	1.7981	0.6141	−1.3596	1.4943	−0.4406	−0.2033	−0.1316
0.8319	0.4270	−0.8888	0.4167	−0.8513	1.1054	1.2237	−0.7003
0.9780	−0.7679	0.8960	0.5154	−0.7165	0.8563	−1.1630	1.8800

1. Random digits imitate values of a random variable with distribution (3.1) (see section 3.1).

2. Normal values imitate values of a normal (gaussian) random variable Z with parameters $a = 0$, $\sigma = 1$.

Bibliography

For further study the reader is referred to the following books. Extensive bibliographies are to be found in the first two listings.

Buslenko, N. P., Golenko, D. I., Sobol', I. M., Sragovich, V. G., and Shreider, Yu. A. *The Monte Carlo Method*. Translated by G. J. Tee. New York: Pergamon Press, 1966. The same work was also published as *The Method of Statistical Testing* (New York: Elsevier Publishing Co., 1964).

Hammersley, J. M., and Handscomb, D. C. *Monte Carlo Methods*. London: Methuen and Co., 1964.

Spanier, Jerome, and Gelband, Ely M. *Monte Carlo Principles and Neutron Transport Problems*. Reading, Mass.: Addison-Wesley Publishing Co., 1969.

Some material dealing with particular matters discussed in chapters 3 and 4 may be found in *Monte Carlo Method* (the proceedings of a symposium), U.S. National Bureau of Standards Applied Mathematics Series, no. 12, 1951.

Tables

Table A. 400 Random Digits[1]

86515	90795	66155	66434	56558	12332	94377	57802
69186	03393	42502	99224	88955	53758	91641	18867
41686	42163	85181	38967	33181	72664	53807	00607
86522	47171	88059	89342	67248	09082	12311	90316
72587	93000	89688	78416	27589	99528	14480	50961
52452	42499	33346	83935	79130	90410	45420	77757
76773	97526	27256	66447	25731	37525	16287	66181
04825	82134	80317	75120	45904	75601	70492	10274
87113	84778	45863	24520	19976	04925	07824	76044
84754	57616	38132	64294	15218	49286	89571	42903

Table B. 88 Normal Values[2]

0.2005	1.1922	−0.0077	0.0348	1.0423	−1.8149	1.1803	0.0033
1.1609	−0.6690	−1.5893	0.5816	1.8818	0.7390	−0.2736	1.0828
0.5864	−0.9245	0.0904	1.5068	−1.1147	0.2776	0.1012	−1.3566
0.1425	−0.2863	1.2809	0.4043	0.6379	−0.4428	−2.3006	−0.6446
0.9516	−1.7708	2.8854	0.4686	1.4664	1.6852	−0.9690	−0.0831
−0.5863	0.8574	−0.5557	0.8115	−0.2676	−1.2496	−1.2125	1.3846
1.1572	0.9990	−0.1032	0.5405	−0.6022	0.0093	0.2119	−1.4647
−0.4428	−0.5564	−0.5098	−1.1929	−0.0572	−0.5061	−0.1557	−1.2384
−0.3924	1.7981	0.6141	−1.3596	1.4943	−0.4406	−0.2033	−0.1316
0.8319	0.4270	−0.8888	0.4167	−0.8513	1.1054	1.2237	−0.7003
0.9780	−0.7679	0.8960	0.5154	−0.7165	0.8563	−1.1630	1.8800

1. Random digits imitate values of a random variable with distribution (3.1) (see section 3.1).

2. Normal values imitate values of a normal (gaussian) random variable Z with parameters $a = 0$, $\sigma = 1$.

Bibliography

For further study the reader is referred to the following books. Extensive bibliographies are to be found in the first two listings.

Buslenko, N. P., Golenko, D. I., Sobol', I. M., Sragovich, V. G., and Shreider, Yu. A. *The Monte Carlo Method*. Translated by G. J. Tee. New York: Pergamon Press, 1966. The same work was also published as *The Method of Statistical Testing* (New York: Elsevier Publishing Co., 1964).

Hammersley, J. M., and Handscomb, D. C. *Monte Carlo Methods*. London: Methuen and Co., 1964.

Spanier, Jerome, and Gelband, Ely M. *Monte Carlo Principles and Neutron Transport Problems*. Reading, Mass.: Addison-Wesley Publishing Co., 1969.

Some material dealing with particular matters discussed in chapters 3 and 4 may be found in *Monte Carlo Method* (the proceedings of a symposium), U.S. National Bureau of Standards Applied Mathematics Series, no. 12, 1951.